GEOGRAPHY NOW!

RIVERS

AROUND THE WORLD

JEN GREEN

Published in 2009 by The Rosen Publishing Group Inc.
29 East 21st Street, New York, NY 10010

First Edition

Editor: Jon Richards
Designer: Ben Ruocco
Consultant: John Williams

Library of Congress Cataloging-in-Publication Data

Green, Jen.
Rivers around the world / Jen Green. – 1st ed.
p. cm. – (Geography now)
Includes index.
ISBN 978-1-4358-2870-4 (library binding)
ISBN 978-1-4358-2956-5 (paperback)
ISBN 978-1-4358-2962-6 (6-pack)
I. Title.
GB1203.8.G738 2009
551.48'3–dc22

2008025744

Manufactured in China

Picture acknowledgments:
(t-top, b-bottom, l-left, r-right, c-center)
Front cover courtesy of NASA, 1 istockphoto.com/Linda Mirro, 4-5 istockphoto.com/Peter Malsbury, 4bl istockphoto.com/Loic Bernard, 5br Dreamstime.com/Sighardo Donnavillaggio, 6-7 istockphoto.com, 6bl courtesy of NASA, 7br istockphoto.com, 8-9 istockphoto.com/Nick Schlax, 8bl istockphoto.com/Vladimir Popovic, 9 courtesy of NASA, 10-11 istockphoto.com/Sandra vom Stein, 10bl istockphoto.com/Paul Tessier, 12-13 istockphoto.com, 13cr istockphoto.com/Linda Mirro, 13br Dreamstime.com/Photowitch, 14-15 istockphoto.com/Nemanja Glumac, 15tr Rabensteiner, 15br Dreamstime.com/Deborah Benbrook, 16-17 Paulo Fridman/Sygma/Corbis, 16bl Dreamstime.com/Ryszard Laskowski, 17br istockphoto.com/Roman Shiyanov, 18-19 istockphoto.com/Luke Daniek, 18bl Dreamstime.com/Alessandro Bolis, 19br Lloyd Cluff/Corbis, 20-21 istockphoto.com, 20bl istockphoto.com/Christopher Steer, 21br Dreamstime.com/Chris Harvey, 22-23 istockphoto.com/Norbert Bieberstein, 23tr Dreamstime.com/Manfred Steinbach, 23br istockphoto.com/Teun van den Dries, 24-25 David Sailors/Corbis, 24bl Dreamstime.com/Wayne Mckown, 25br Robert Holmes/Corbis, 26-27 istockphoto.com/Valerie Crafter, 26bl courtesy of NASA, 27br Dreamstime.com/Aschwin Prein, 28-29 Bob Krist/Corbis, 28bl courtesy of NASA, 29tr Reuters/Corbis

CONTENTS

What are rivers? 4
A river's journey 6
Shaping the landscape 8
Rivers as habitats 10
Using rivers 12
Protecting rivers 14
Earth's greatest river 16
Watering the desert 18
London's historic river 20
Europe's busiest river 22
Taming America's giant 24
Harvesting China's river 26
Asia's sacred river 28
Glossary, Further Information, and Web Sites 30
Rivers topic web 31
Index 32

What are rivers?

Rivers are flows of fresh water running through the landscape. Most rivers start in hills or mountains. Pulled by gravity, they head downhill to empty into a sea, lake, swamp, or another river. Rivers are important habitats for wildlife, and are vital to people, too.

NATURAL PROCESSES

Rivers play a major role in the water cycle. They return rainwater to seas and lakes, from where it rises up into the air as water vapor and later condenses to form rain clouds. Less than half of all rainwater drains into rivers, however—the rest is absorbed by plants or seeps into the ground. Rivers also help to shape the landscape, carrying away rocks and soil, and dropping them to form new land downstream.

Many of the world's largest cities have grown up on rivers or at river mouths. This is Shanghai, near the mouth of the Yangtze River in China.

USEFUL RIVERS

Rivers support many kinds of wildlife, including fish, frogs, birds, and insects. Drinking water is drawn from rivers, or from reservoirs made by damming rivers. Rivers are also used in farming, industry, and transportation. Unfortunately, people also use rivers to dump waste. This creates the need to clean up and protect rivers.

In hot regions, some rivers dry up before they reach the ocean. The Okavango River in southern Africa dwindles and dries up as it enters the Kalahari Desert, forming a swamp.

Changing rivers

Rivers change with the seasons. Some are frozen in the winter, and many are swollen by rain and melted snow in the spring. In hot countries, some rivers dry up completely in the summer. People also make major changes to rivers—by building dams or digging canals, for example.

Some rivers freeze over in the winter. This family is skating on the River Danube near Vienna in Austria.

A river's journey

Rivers pass through several stages on their journey to the ocean. There are three main stages in a river's course: the upper, middle, and lower course. These stages are also described as the river's youth, middle age, and maturity or old age.

From above, most river systems resemble a tree with many branches. This aerial view shows the muddy waters of the Solim es River joining the Amazon in South America.

UPPER COURSE

Rivers generally begin life in uplands, where they are fed by rainwater, streams, springs, or ice from glaciers. From the source (beginning), young rivers flow swiftly downhill through steeply sloping ground. White-water rapids form where the river tumbles over boulders in a shallow bed. As more streams and minor rivers, called tributaries, join the main river, so the volume of water grows.

The Iguaçu Falls are a spectacular set of falls in South America. They lie on the border between Brazil and Argentina.

LOWER COURSE

In middle age, the river flows through gently sloping ground. From a wide valley, it enters a broad plain called a floodplain (because of the risk of flooding after heavy rain). As the mature river approaches the sea, it flows through a wide channel called an estuary. Seawater mingles with freshwater here at high tide. The river and all its tributaries form a river system.

Waterfalls

Waterfalls form where rivers flow from hard rock on to softer rock. The soft rocks wear away more quickly to make a ledge, over which the water cascades. Most waterfalls form on the upper courses of rivers. The world's widest falls, on the River Khone in Laos, Southeast Asia, are 6.7 miles (10.8 km) wide.

Angel Falls in Venezuela is the world's highest waterfall. Here, the River Churún plunges 3,212 ft. (979 m) as it cascades down from a plateau.

Shaping the landscape

Rivers are the main force at work shaping the landscape. As they rush downhill, they wear away rocks and soil in a process called erosion. Rocky debris bouncing along the bed—known as the river's load—increases the river's cutting power.

In the United States, the Colorado River has cut down through a high plateau to form the majestic Grand Canyon—one of the world's deepest canyons at 1 mile (1.6 km) deep.

Water swinging around a bend flows fastest on the outside, eating away at the banks. Sediment is dropped on the inside, where the current is slower. Over time, curves deepen to become loops called meanders.

EROSION

Rivers have most power to erode the landscape where they flow over soft rocks or down steep terrain (which increases the speed of flow). Rocks carried by the foaming water are gradually smashed to form fine debris called sediment. Young rivers erode deep, V-shaped valleys as they flow through hills. They can carve sheer-sided gorges where they slowly cut down through hard rock.

DEPOSITION

As middle-aged rivers leave the hills, they begin to flow more smoothly. The river starts to drop its load, in a process called deposition. Large boulders are dropped first, then fine debris. As they flow through broad valleys, rivers weave from side to side, forming curves that gradually deepen as erosion continues. Near the river mouth, the river drops its sediment to form mudflats and deltas, with islands and spits farther out to sea.

Deltas

Deltas are large expanses of marshy ground at river mouths. They are made of fine silt dropped by the river as it enters the sea, where the current slows abruptly. The river often separates into channels called distributaries as it winds through the delta.

Some deltas are fan-shaped. Others, such as the Mississippi Delta, resemble a bird's foot when viewed from above.

Rivers as habitats

All living things need water. Rivers provide habitats for wildlife, ranging from insects, worms, and mollusks, to frogs, birds, and mammals. Fish, snails, and worms live in the river, while bears, otters, and alligators hunt along the banks. Frogs, dragonflies, and mosquitoes visit rivers to breed.

CLIMATE AND CONDITIONS

A river's climate mainly depends on its location in a tropical, temperate, or cold region. Some river habitats are linked to a particular climate—mangrove forests, for example, grow only on estuaries in the tropics. Altitude (height) can also affect climate, since many rivers begin in snowy mountains. Rivers offer different habitats along their course, from rushing streams near the source, to wooded valleys lower down and mudflats by the sea.

Alligators are among the largest predators that hunt by rivers. These reptiles prey on fish and turtles, and will also seize animals that come to drink.

In North America, brown bears have learned the breeding habits of salmon. The bears prey on the fish as they swim upriver to spawn.

AQUATIC LIFESTYLES

River animals are suited to their aquatic lifestyle. The powerful tails of beavers, fish, and river dolphins make them strong swimmers. Swans, geese, and otters have webbed feet. Some creatures live only in one part of a river. Birds called dippers inhabit young rivers, for example, kingfishers and otters hunt along middle-aged rivers, and wading birds feed in muddy estuaries.

Migration

Some animals use rivers as watery highways to guide them during seasonal migrations. Salmon hatch in streams, then swim downriver to mature in the sea. They swim back upriver to breed. Birds fly along rivers as they migrate between their summer breeding and winter feeding grounds.

The Mississippi River guides millions of ducks, geese, and other birds as they move north and south.

Using rivers

Fresh water is vital to people, as well as plants and animals. Rivers are useful to people in a number of ways, providing water for drinking, farming, industry, energy, and transportation. The world's earliest civilizations grew up along rivers.

FARMING AND SETTLEMENTS

River water has been used for irrigation since ancient times. Land next to rivers is particularly suited to farming because of the rich silt that is deposited when the river floods. Rivers act as highways for shipping, and roads are often built along river valleys. In early times, villages and towns were often established at confluences where two rivers met.

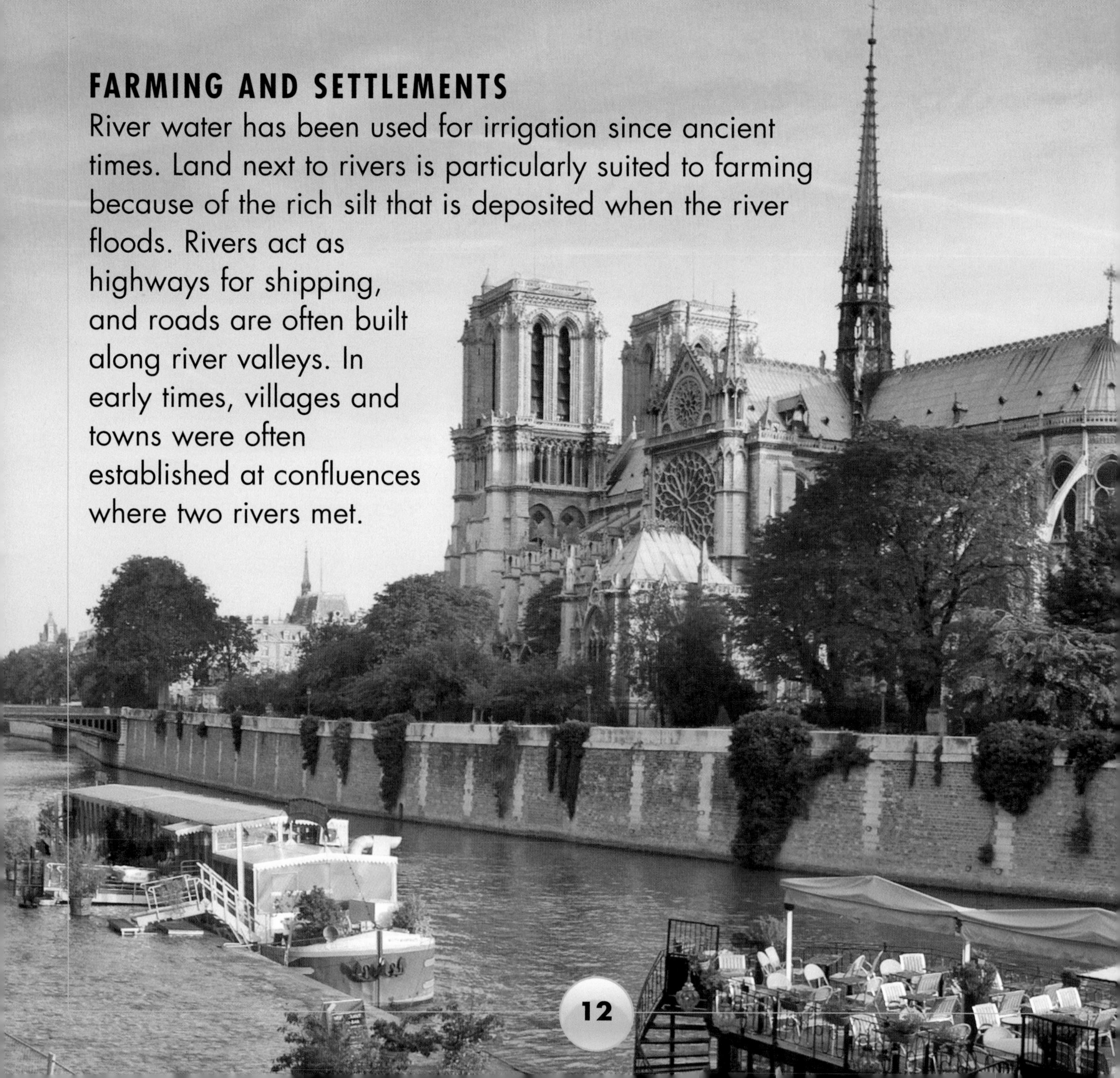

One of the earliest uses of rivers was for defense. Settlements grew up on islands, or where a meander offered protection from attack. Notre Dame Cathedral in the heart of Paris was built on an island in the River Seine.

MINING AND MANUFACTURING

Rivers provide gravel and sand for building, and occasionally, valuable minerals such as iron, gold, and diamonds. The energy of flowing water can be harnessed to produce electricity. All kinds of factories, from paper mills to chemical plants, use river water for cooling, cleaning, and manufacturing.

Rivers are used for many leisure activities, such as fishing and sailing. This angler (right) is fly-fishing on a river in Colorado.

Transportation

Rivers provide a cheap means of transporting goods such as grain, coal, and building stone. The river's course is often altered to make it easier for boats to navigate. Shallow parts are dredged to make the water deeper. Weirs and locks allow boats to pass obstacles such as rapids.

Canals are often built to straighten a river's course or to link two rivers. This houseboat is on a canal in France.

Protecting rivers

Towns, farms, and factories produce pollution that can harm river life—and the people who live nearby. Rivers are also changed by the construction of dams, canals, and reservoirs. There is a constant need for conservation work on rivers worldwide.

Plastic bottles and green slime on this river are evidence of pollution. The green coating is caused by algae (tiny plants) breeding in polluted water. Algae reduce oxygen levels in the water, which can kill fish.

POLLUTION

For hundreds of years, riverside towns flushed sewage and other waste into their local rivers. Nowadays, in developed countries, sewage must be treated before used water is returned to the river. Local farms use fertilizers and pesticides to improve crop yields—these chemicals can harm river life, as can waste from mines and factories. Many countries now have strict laws controlling the dumping of waste in rivers.

HUNTING AND HABITATS

The construction of new towns and factories on rivers nibbles away at wild river habitats. Conservation measures include the setting aside of wild rivers as reserves, or tightening hunting laws to protect river species that are at risk of dying out.

In the U.K., water voles are now very rare because of habitat loss. New predators, such as American mink, also prey on these animals. Conservation work includes the construction of new wetland reserves.

New species

River habitats can be harmed by the introduction of plants and animals that do not belong there. In the 1900s, water hyacinth plants were introduced to many river systems, because their blue flowers were considered attractive. These fast-growing plants now choke rivers, leaving no room for native plantlife.

Water hyacinths can make rivers impassable for boats. These weeds are almost impossible to get rid of once they become established.

Earth's greatest river

The Amazon

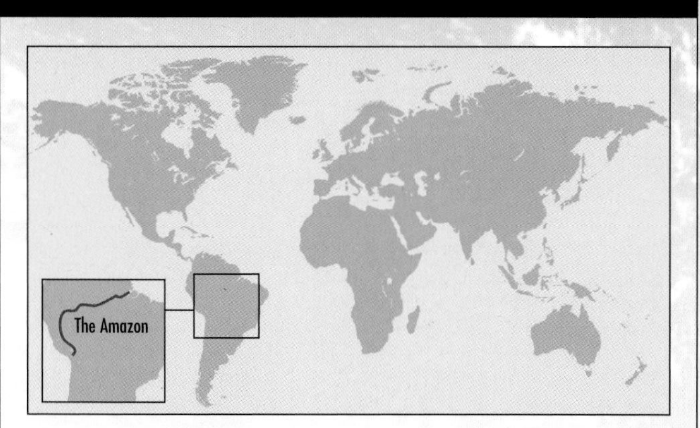

STATISTICS

- *Location: Northern South America*
- *Length: 4,002 miles (6,440 km)*
- *Notable features: Vast drainage basin, 1,100 tributaries*
- *Major industries: Forestry, farming, mining*
- *Environmental issues: Habitat loss caused by deforestation; pollution and erosion from mining and towns*

The Amazon contains a much greater volume of water than any other river, carrying a fifth of the world's river water. It is the world's second-longest river, flowing for 4,002 miles (6,440 km). As it passes through the Amazon Basin—the world's largest rain forest—it is joined by around 1,100 tributaries, many of which are major rivers in their own right.

AMAZON BASIN

A river's drainage basin is the total area drained by a river and its tributaries. The Amazon has the world's largest river basin, draining much of northern South America. The river is home to a huge variety of plants and animals, including many giant species, such as enormous water lilies and the arapaima—the world's largest freshwater fish.

Giant species of the Amazon include the capybara (left), the world's largest rodent, and the anaconda, one of the world's biggest snakes.

Annual floods cover the forest in water up to 33 ft. (10 m) deep. This flooding can last for six months of the year.

CHANGING HABITAT

The Amazon floods every year in the rainy season, swollen by rain and melted snow from the Andes Mountains. The floods transform the forest, with fish and alligators swimming among trees that used to be home to parrots and monkeys. People have also brought many changes to the Amazon in the last 50 years, cutting down large areas of forest for timber and to make room for new farms, mines, and towns.

Amazon peoples

Amerindian tribes such as the Yanomami have lived along the Amazon for centuries. Forest people traditionally lived by hunting, fishing, and growing crops, such as sweet potato and cassava. In recent years, tribal lands have been whittled away by mines, dams and new settlements. Local ways are changing fast.

Amerindian tribal groups are now fighting to save the forests that have sustained them for generations.

Watering the desert

The Nile

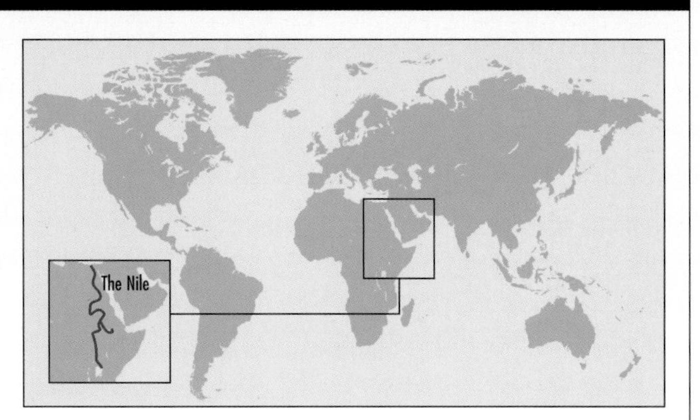

STATISTICS

- *Location: Northeast Africa, including Ethiopia and Egypt*
- *Length: 4,160 miles (6,695 km)*
- *Notable features: Aswan Dam, Lake Nasser, triangular delta*
- *Major industries: Farming, hydroelectricity, tourism*
- *Environmental issues: Pollution; Aswan Dam prevents flooding and reduces fertility of farmland*

The Nile is the world's longest river, flowing for 4,160 miles (6,695 km). It is formed by the White Nile and Blue Nile, which rise in the mountains of east Africa. On its long course, the river flows north through mountains, gorges, swamps, and a vast desert. A large delta has formed where it drains into the Mediterranean Sea.

Ancient monuments, such as temples and the pyramids, make tourism a major industry along the Nile. This is the temple of Abu Simbel near Lake Nasser.

FLOODS AND FERTILITY

The rainy season in the Ethiopian Highlands causes the Nile to flood, particularly along its lower course in Egypt. Floodwaters spread rich silt across the valley, transforming a strip of desert into green, fertile land. Farmers cultivated crops here in ancient times, giving rise to one of the world's earliest civilizations. The river has been used for food, farming, and transportation ever since, and in recent times, for industry and energy.

THE ASWAN DAM

In 1970, an enormous dam was completed at Aswan in southern Egypt to generate electricity. The dam formed Lake Nasser, a vast lake providing water for drinking and farming. However, the dam also prevents the annual flood, which has caused problems. Farmers now have to use artificial fertilizers to replace silt once deposited by the river.

Fields along the Nile are irrigated using a network of canals and ditches. Farmers here can harvest two or three crops a year.

Hydroelectric power

Hydroelectric power stations harness the energy of flowing water to generate electricity for homes and industry. A river is usually dammed above the power station, so rushing water flows past turbines. The turbines spin to operate generators that produce electricity. Power lines take the electricity to distant cities.

As water is released through the Aswan Dam hydroelectric power station, it produces electricity for much of Egypt.

London's historic river

The Thames

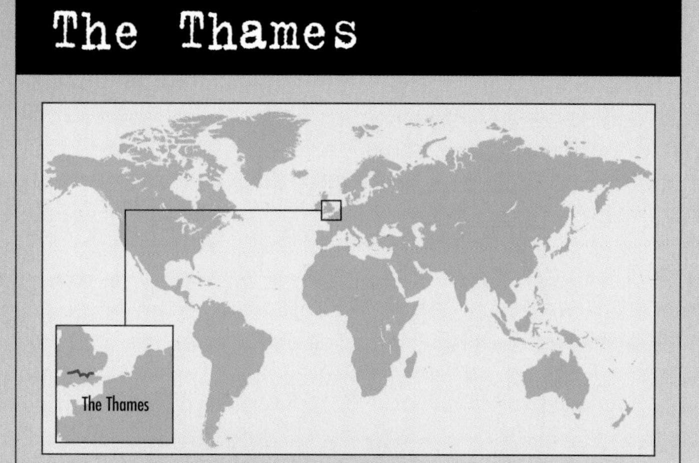

STATISTICS

- *Location: Southeast United Kingdom*
- *Length: 215 miles (346 km)*
- *Notable features: Thames Barrier, wide estuary*
- *Major industries: Farming, manufacturing in Thames Valley*
- *Environmental issues: Pollution from industry and towns is curbed by law*

The Thames is the U.K.'s second-longest river after the Severn. It rises in the Cotswold Hills in western England and flows through London, before draining into the North Sea.

The London Eye or Millennium Wheel is one of the river's new attractions. The Houses of Parliament, seen in the background, are probably the most famous sight.

A MAJOR PORT

London was founded by the Romans in about 50 CE, at a crossing point on the Thames. Located at the highest point that could be reached by ocean-going ships, it expanded into a major port. However, by the mid-1900s, ships had become too large to enter London, and the docks are no longer used.

The Tower of London is one of the city's most important historic attractions, housing the Crown Jewels.

TOURISM AND REGENERATION

Tourism is a now major industry in London. The Thames is lined with historic buildings and new attractions, such as the Tate Modern art gallery and the rebuilt Globe Theatre. London's Docklands were transformed in the 1980s, when old warehouses and factories were replaced by office buildings, houses, and stores.

The Thames Barrier

During very high tides and storms, the East End of London is at risk from flooding. The Thames Barrier was completed in 1982 to protect the capital from floods. Gates can be raised to block tidewater.

The floodgates of the Thames Barrier normally lie on the river bed, allowing ships to pass upriver.

Europe's busiest river

The Rhine

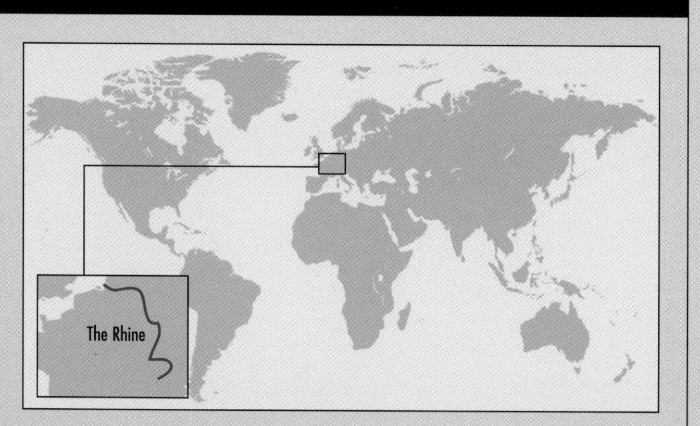

STATISTICS

- *Location: Western Europe*
- *Length: 820 miles (1,320 km)*
- *Notable features: Narrow gorge in Germany; large delta at mouth*
- *Major industries: Manufacturing, farming*
- *Environmental issues: Pollution from industry and towns*

The Rhine is one of Europe's longest rivers. It begins high in the Swiss Alps and flows down into France and Germany, through gorges and then a wide plain. When it reaches the Netherlands, it drains into the North Sea. Aquatic life in this busy industrial river has been harmed by pollution.

TRANSPORTATION HIGHWAY

The Rhine is navigable for 500 miles (800 km), thanks to regular dredging. Oceangoing ships can travel inland as far as Cologne in Germany, and large barges can pass to the Swiss port of Basel. This has made Basel the biggest port in landlocked Switzerland. A string of castles along the Rhine, as it flows through a narrow gorge in Germany, shows the river's strategic importance since medieval times.

POLLUTION FROM INDUSTRY

The Rhine links with the Ruhr River in Germany's industrial heartland. Factories line the river here, and along its course. In 1986, a fire at a Swiss chemical plant released 30 tons of poisonous waste, including mercury and pesticides, into the river. All aquatic life was poisoned for 62 miles (100 km) downstream. This was a major setback for the Rhine countries' efforts to reduce pollution.

Chemical plants along the Rhine have to follow strict antipollution guidelines.

Food, minerals, iron, and steel are transported along the Rhine using barges like these (right).

Sharing river resources

Pollution becomes a particularly sensitive issue in cases where a river flows through more than one country. If the upper course of the river is polluted, countries downriver will suffer the consequences. Sharing river water equally is also important, especially in hot, dry regions where irrigation is vital to farming.

Towns along the lower Rhine can be badly affected by pollution upriver. This is the Dutch port of Rotterdam, the last city on the Rhine, 31 miles (50 km) from the sea.

Taming America's giant

The Mississippi

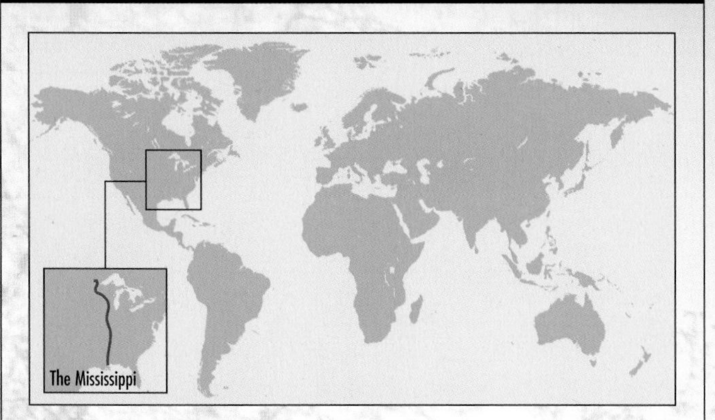

STATISTICS

- *Location: United States*
- *Length: 2,340 miles (3,765 km)*
- *Notable features: Swamps and birdsfoot delta at mouth*
- *Major industries: Manufacturing, farming, hydroelectricity*
- *Environmental issues: Pollution from industry and farming; flood risk*

The Mississippi, the world's fourth-longest river, flows for 2,340 miles (3,765 km) and empties into the Gulf of Mexico near New Orleans. Together with tributaries such as the Missouri, the Mississippi drains over half of North America.

The Mississippi has been a major highway for shipping since the days of steamboats in the 1800s. Millions of tons of freight, including grain, cotton, and coal, are still transported by barge.

MISSISSIPPI FLOODPLAIN

In its middle course, the Mississippi flows through a wide floodplain made fertile by river silt. Over thousands of years, regular floods have dropped earth and gravel by the river to form high natural banks called levées. In many places, the river's channel is higher than the surrounding plain. This can cause floods. In 1993, the Mississippi flooded 17,000 sq miles (44,000 sq km) of land.

FLOODS IN NEW ORLEANS

In August 2005, New Orleans suffered a catastrophic flood in the wake of torrential rain and high seas whipped up by Hurricane Katrina. Four-fifths of New Orleans was flooded, and the entire population had to be evacuated. Over 1,000 people died, and it was many months before city-dwellers could return.

Floodwaters cover a small town by the Mississippi in 1993. The water drained away to leave 70,000 homes polluted by mud.

Flood defenses

Dams and reservoirs along the Mississippi help to prevent flooding. Channels called spillways take excess water, and the river is also dredged regularly. Along many stretches, the height of the levées has been raised using concrete walls. If these artificial defenses are breached, as they were in 2005, devastating floods can result.

Artificial levées keep floodwaters at bay. Walls up to 53 ft. (16 m) high protect cities such as St. Louis and New Orleans.

Harnessing China's river

The Yangtze

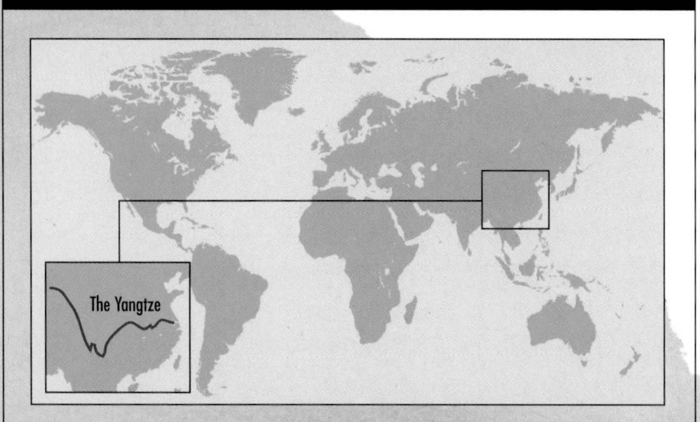

STATISTICS

- *Location: Central and eastern China*
- *Length: 3,915 miles (6,300 km)*
- *Notable features: Narrow gorges, floodplain, delta at mouth*
- *Major industries: Manufacturing, farming, hydroelectricity*
- *Environmental issues: Pollution from industry, towns and farming; dam alters river habitats*

The Yangtze is the world's third-longest river. It rises in the mountains of Tibet and flows eastward to reach the East China Sea near Shanghai. On its way, it is joined by 700 tributaries that swell its muddy waters. Its Chinese name, Chang Jiang, simply means "the long river."

As it passes through the Three Gorges region, the Yangtze is 656 ft. (200 m) deep—the world's deepest river. The construction of the dam has harmed the river's scenic beauty.

This satellite view (below) shows the 404-mile (650-km) reservoir that has formed behind the Three Gorges Dam, flooding the narrow river valley.

CENTER FOR CIVILIZATION

Like the Nile, the Yangtze was an early center of civilization, with major canal works dating as far back as 500 BCE. Now, about half of China's huge population lives on or near the river. The Yangtze has been used for transportation and irrigation for thousands of years, but it also has a history of flooding.

DAM DISPUTE

The Three Gorges Dam, located in a narrow gorge midway along the Yangtze's course, has caused controversy because of its impact on the environment. Several cities have been flooded by the huge lake that has formed behind the dam, and 1.2 million people have been relocated.

The Three Gorges project

Construction of the Three Gorges Dam began in 1994. The scheme is expected to be fully operational by 2011. A series of locks allows ships to pass upriver, but the dam also forms a barrier for migrating wildlife. Conservationists fear it may endanger rare species, such as the Yangtze alligator and river dolphin.

The Three Gorges project has involved the construction of a barrier 607 ft. (185 m) high and 1.24 miles (2 km) wide. The dam will supply 10 percent of China's energy.

Asia's sacred river

The Ganges is one of the greatest rivers in southern Asia. Its source is a glacier high in the Himalayas. From the mountains, the Ganges flows east across the north Indian plain into low-lying Bangladesh. Here, it joins the Brahmaputra River to form the world's largest delta.

The Ganges

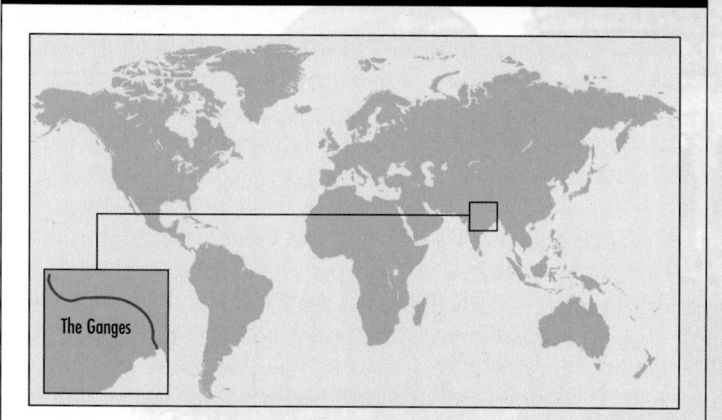

STATISTICS

- *Location: India, Bangladesh*
- *Length: 1,560 miles (2,510 km)*
- *Notable features: Swamps and huge delta at mouth*
- *Major industries: Farming, fishing, some manufacturing*
- *Environmental issues: Pollution from settlements and farming; flood risk*

HOLY RIVER

The Ganges Basin covers about 386,102 sq miles (1 million sq km). It is home to about 400 million people, which makes it the most densely populated river on Earth. The Ganges is sacred to Hindus, who call it Ganga Mai—"Mother Ganges." Every year, about a million Hindu pilgrims journey to the holy cities of Varanasi and Allahabad to bathe in the Ganges, which is believed to wash away sins.

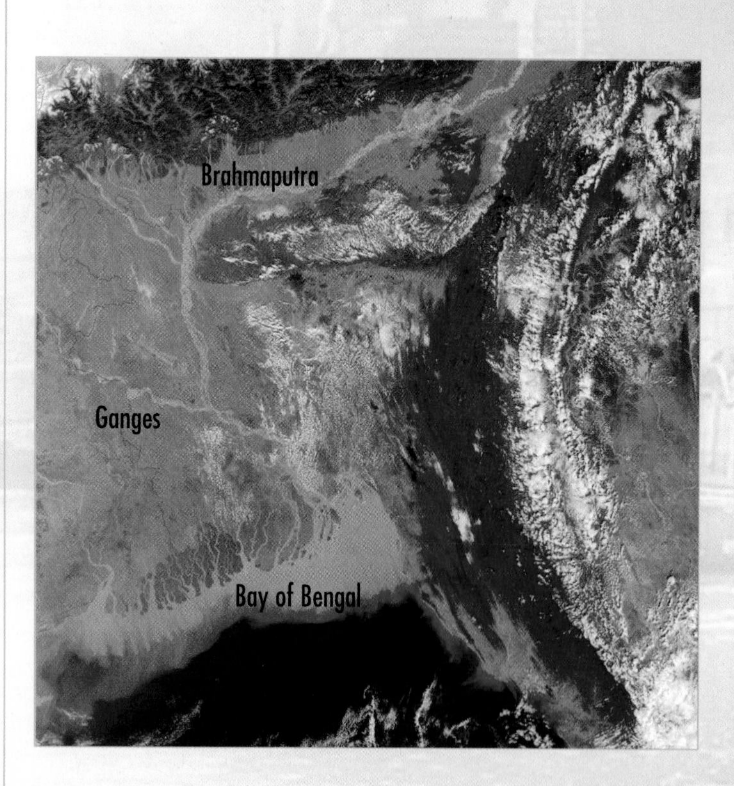

A satellite view shows the silt-laden Ganges and Brahmaputra rivers flowing through a green floodplain as they approach their delta on the Bay of Bengal.

Coastal floods

Hurricanes are another threat to Bangladesh, bringing severe coastal flooding. The flood risk is greatest as the hurricane sweeps ashore from the ocean, bringing a huge wall of water called a storm surge along with it. In 1991, 138,000 Bangladeshis died when a hurricane struck the delta.

People wade through waist-high water during a flood in Bangladesh's Kushtia district in 2000.

GANGES DELTA

The delta of the Ganges and Brahmaputra Rivers covers 29,000 sq miles (75,000 sq km), mostly in Bangladesh. Yearly floods cover the delta with rich silt, but they can bring misery as well as fertile soil. In 1998, floodwaters covered nearly 70 percent of Bangladesh and 25 million people were made homeless.

Pilgrims bathe in the Ganges at Varanasi. Hindus also scatter the ashes of their dead relatives in the river.

Glossary, Further Information, and Web Sites

AQUATIC
Of living things that dwell in water.

CANAL
An artificial waterway.

CONDENSE
When water changes from a gas into a liquid.

CONFLUENCE
The point where two rivers merge.

DELTA
A flat, swampy area made of sediment dropped by a river at its mouth.

DEPOSITION
When a river drops its load of rock and sediment.

DISTRIBUTARY
A minor channel of a river.

DRAINAGE BASIN
The total area drained by a river and its many tributaries.

DREDGE
To deepen a river by removing rock and sediment from its bed.

EROSION
When rock or soil is worn away by water, ice, or wind.

ESTUARY
The mouth or lower stretch of a river, washed by seawater at high tide.

EVACUATE
When everyone is told to leave an area because of danger, such as fire or flood.

HYDROELECTRICITY
Electricity generated using the energy of fast-flowing water.

IRRIGATION
When farmers water their fields in order to grow crops.

LEVÉE
A raised bank edging a river, made of sediment dropped by the river in flood.

LOAD
The rocky debris carried along by a river, including boulders, stones, sand, and mud.

MIGRATION
A seasonal journey made by an animal species to find food, shelter, or a safe place to breed.

MUDFLAT
A low-lying bank located near the mouth of a river.

RESERVOIR
An artificial lake used to store water.

SEDIMENT
Fine rocky debris, such as sand or mud.

SEWAGE
Dirty water from homes, containing human waste.

SILT
Fine pieces of rock that have been ground down to form clay, sand, or mud.

SOURCE
The place where a river begins, such as a spring or a lake.

SPAWN
When fish or frogs breed by laying their eggs in water.

SPILLWAY
A channel designed to drain excess water from a river.

SPIT
A narrow piece of land stretching into the sea.

TRIBUTARY
A stream or small river that joins a bigger river.

TURBINE
A machine where steam, gas, or water turn a bladed wheel in order to generate electricity.

FURTHER READING

Geography Fact Files: Rivers
by Mandy Ross
(Smart Apple Media, 2004)

Precious Earth: Saving Oceans and Wetlands
by Jen Green
(Chrysalis Education, 2004)

WEB SITES

Due to the changing nature of Internet links, PowerKids Press has developed an online list of Web sites related to the subject of this book. This site is updated regularly. Please use this link to access this list:
www.powerkidslinks.com/geon/river

Rivers topic web

Use this topic web to discover themes and ideas in subjects that are related to rivers.

RIVERS

GEOGRAPHY

- Stages of a river.
- Formation features such as gorges, waterfalls, and deltas.
- How rivers shape the landscape through erosion and deposition.
- Use of rivers, including transportation, irrigation, industry, and energy.

SCIENCE AND THE ENVIRONMENT

- Rivers as habitats for wildlife.
- Role of rivers in the water cycle.
- Causes of river floods, and flood control.
- Altering the course of rivers, for example, canal construction.
- Environmental problems such as pollution.
- Conservation work to tackle environmental problems on rivers.

ART AND CULTURE

- Art and culture of people who live on or near rivers.
- Myths and legends that contain or are inspired by rivers.
- Music and artworks that feature or are inspired by rivers.

ENGLISH AND LITERACY

- Stories and accounts of the lives of river people, such as *The Adventures of Tom Sawyer* by Mark Twain.
- Debate the pros and cons of dam construction on rivers.

HISTORY AND ECONOMICS

- Development of settlements and industries next to rivers.
- Trade routes along rivers.

Index

A, B

alligators 10, 17, 27
Amazon river 6, 16–17
Aswan Dam 19
bears 11
birds 11

C, D

canals 5, 13, 14, 19, 27
chemicals 14, 23
conservation 5, 14–15
dams 5, 14, 17, 19, 25, 27
deltas 9, 18, 29
deposition 4, 9
dredging 13, 22, 25

E

Egypt 18–19
electricity 13, 19
energy 12, 13, 19
environment ...14–15, 23, 27
erosion 4, 8–9
estuaries 7, 10

F, G

factories 13, 14, 23
farming 5, 12, 14, 19
fertilizers 14, 19
fish 16, 17
flood defenses 21, 25
floodplains 7, 25, 28
floods 17, 19, 25, 27, 29
Ganges river 28–29
gorges 9
Grand Canyon 8

H, I, L

habitats 10, 15, 17
hurricanes 25, 29
hydroelectric power 19
industry 5, 12, 19, 23
irrigation 12, 19, 23, 27
leisure activities 13
London 20–21
lower course 6, 7

M

mangrove forests 10
marshes 10
meanders 8
middle course 6, 7
migration 11, 27
mining 13, 17
Mississippi 9, 11, 24–25
mountains 10
mudflats 9, 10

N, P

Nasser, Lake 18, 19
New Orleans 25
Nile river 18–19
plants 15, 16
pollution 5, 14, 23
ports 21, 22

R

rainwater 4, 7, 17
rapids 7, 13
reservoirs 5, 14, 25, 26
Rhine river 22–23
river system 7
rocks 8, 9

S

salmon 11
sediment 9
settlements 4, 12, 17
sewage 14
silt 9, 12, 19, 25, 29
streams 7, 10

T

Thames river 20–21
Three Gorges Dam 26, 27
tourism 18, 21
towns and cities 4, 14, 15
transportation ..12, 13, 19, 22, 23, 24, 27
tributaries 7, 16, 24

V, W, Y

valleys 9, 10, 12
water cycle 4
waterfalls 7
water voles 15
Yangtze river 4, 26–27